A Practical Guide to
Gender Diversity for
Computer Science Faculty

Synthesis Lectures on Professionalism and Career Advancement for Scientists and Engineers

Editor
Charles X. Ling, *University of Western Ontario*
Qiang Yang, *Hong Kong University of Science and Technology*

Synthesis Lectures on Professionalism and Career Advancement for Scientists and Engineers includes short publications that will help students, young researchers, and faculty become successful in their research careers. Topics include those that help with career advancement, such as writing grant proposals; presenting papers at conferences and in journals; social networking and giving better presentations; securing a research grant and contract; starting a company, and getting a Masters' or PhD degree. In addition, the series will publish lectures that help new researchers and administrators to do their jobs well, such as: how to teach and mentor, how to encourage gender diversity, and communication.

A Practical Guide to Gender Diversity for Computer Science Faculty
Diana Franklin
2013

The Handbook of Analytical Writing for Science, Engineering, and Business
William E. Winner
2013

A Practical Guide to Gender Diversity for Computer Science Faculty

Diana Franklin

ISBN: 978-3-031-01380-5   paperback
ISBN: 978-3-031-02508-2   ebook

DOI 10.1007/978-3-031-02508-2

A Publication in the Springer series
SYNTHESIS LECTURES ON PROFESSIONALISM AND CAREER ADVANCEMENT FOR SCIENTISTS AND ENGINEERS

Lecture #2
Series Editors: Charles X. Ling, *University of Western Ontario*
              Qiang Yang, *Hong Kong University of Science and Technology*
Series ISSN
Synthesis Lectures on Professionalism and Career Advancement for Scientists and Engineers
ISSN pending.

A Practical Guide to Gender Diversity for Computer Science Faculty

Diana Franklin
University of California, Santa Barbara

SYNTHESIS LECTURES ON PROFESSIONALISM AND CAREER ADVANCEMENT FOR SCIENTISTS AND ENGINEERS #2

ABSTRACT

Computer science faces a continuing crisis in the lack of females pursuing and succeeding in the field. Companies may suffer due to reduced product quality, students suffer because educators have failed to adjust to diverse populations, and future generations suffer due to a lack of role models and continued challenges in the environment. In this book, we draw on the latest research in sociology, psychology, and education to first identify why we should be striving for gender diversity (beyond social justice), refuting misconceptions about the differing potentials between females and males. We then provide a set of practical types (with brief motivations) for improving your work with undergraduates taking your courses. This is followed by in-depth discussion of the research behind the tips, presenting obstacles that females face in a number of areas. Finally, we provide tips for advising undergraduate independent projects or graduate students, supporting female faculty, and initiatives requiring action at the institutional level (department or above).

KEYWORDS

diversity, gender, advising, education

To my daughter, Sara,
who inspires me to effect change on a global scale
so that the world will be more gender-blind
by the time she reaches college.

Contents

Acknowledgments

I could not have written this book without the existence of NCWIT, the National Consortium of Women in Technology. Through annual summits and online resources, NCWIT gathers the latest research relevant to the reasons why females are such a small minority, in computing the barriers they face, and the techniques that work. This research is disseminated in the summit plenary talks, small sessions, discussion groups, and panels. In addition, resources on line vary between complete in-a-box guides, brief test cases, and top-10 tips. I have gained knowledge and, perhaps more important, inspiration to share what I have learned there, to delve deeper into the issues raised there through further research, and to write a book that adds to this growing variety of ways to convey this very important message: We have a major problem, we have insights as to what may be the causes, and there is something you can do to help.

I also want to thank my amazing husband, Fred Chong, for not only believing in gender equity, but following through with actions. Instead of viewing my career either less important than or in competition with his, we have always worked together as a team to find solutions so we can both achieve our professional goals.

Finally, I want to thank my family for indulging and encouraging my belief that I could do anything my big brothers could do.

Diana Franklin
April 2013

CHAPTER 1

Introduction

In the last 20 years, the female representation in computer science has continued to drop, despite substantial time and effort to change this. Part of the reason it has failed is that, while we are better understanding why females are not well represented, we are far behind in implementing changes. While computer science professors are powerless to change the first 18 years of a student's life, there are many things one can do for undergraduates, graduate students, and faculty to even the playing field. Many professors are unaware how big of a difference they can make—and that many of their current practices may contribute to the negative environment for females (or, worse, be instances of unintentional discrimination). The high-level goal of this book is to provide a short, concise guide for computer science professors to understand the background of the problem, what changes they can make, and the research behind those changes.

In summary, first, we want to provide computer science professors concise explanations of the continuing societal influences that provide different environments for males and females. This is both to provide information for those already on board with diversity efforts and to sway those who are skeptical about the need for interventions. Second, we provide practical tips for computer science professors so they can make positive changes in the classroom for the masses, in one-on-one advising, and in supporting faculty. Finally, we provide the research on which these tips are based in order to show that these suggestions are grounded in research as well.

This book is aimed at a United States audience, because the root causes are cultural. Many different cultures have some overlap in the environment in which women are raised, so some of these lessons can be applied to different countries. But the research we use is largely based on Americans, so we cannot make any statements about other countries.

In addition, because this book is for computer science professors, we focus on the things that they can accomplish. There are many large cultural norms that contribute to this problem, but solutions are far beyond the scope of this book. We raise these as issues, but only from the perspective of how they affect the students who arrive at university. The solutions we pose are ones that can be instituted at a university.

Increasing diversity requires us to delve into why the women who *could have been* computer scientists *are not* computer scientists. This is actually an impossible question, since we cannot know which ones would have chosen it and been successful in a different environment. If we ask about the experiences of the females in computer science, we are unlikely to find the answer, because these women had either the personality, upbringing, or experience in computer science that allowed them to not just persevere, but thrive. Therefore, the data in this book sometimes does not match my experience, nor those of other female computer scientists. Instead, this book represents a data-driven perspective on diversity rather than anecdotes.

The evidence in this book is based on statistical observations of human behavior, sometimes recorded by an observer, other times in a controlled environment. We then take these conclusions and apply them to the computer science environment, as experienced by females. Just because the conclusions are based on studies does not mean that the conclusions always hold, because psychological studies typically point out correlations rather than causal relationships. We then take several studies that, taken together, could reasonably conclude some causal relationship.

In addition, no matter what the statistics say about the group, there is often more variation within the group than across groups. Some females will identify with the "male" patterns, and some males will identify with the "female" patterns. The purpose is to identify how to best attract a large segment of the population.

Finally, behaviors that are a product of society change through time. What was true in this generation may not be true in the next. If we are aware of these societal effects, we can both develop the students who were influenced by them and change how we treat young women in order to prevent them from experiencing the same fate.

Chapter 2 makes the case for actively pursuing gender diversity in computer science. If you are reading this for the tips because you are already on board with the premise, then we encourage you to skim through the graphs and pictures to reinforce your existing beliefs that females have equal capability but not yet equal opportunity to pursue computer science. Most people we encounter are intrigued by the data. If you are a skeptic, then please read the chapter, because we believe the evidence for active intervention is compelling.

Chapter 3 then presents different obstacles females face, along with the research behind them and explanations of how this relates to computer science.

Chapter 4 provides concrete tips along with brief motivations for how faculty can structure their classes and interactions with all students so that students who have similar attributes to the average female (i.e., low confidence, less experience, minority) can more likely succeed. The hope is that these tips and brief explanations will motivate you to read more about the reasons behind them.

Finally, we provide three more chapters giving practical tips for advising females, supporting female faculty, as well as changes that require institutional buy-in (either by the department or university).

At the end of each chapter, for those who are particularly interested in one of the areas we discuss, we provide resources that you can read that provide more details about subjects we talked about in the chapter. Like this book, these resources are not the original works, but instead reports and guides that summarize information and provide suggestions. There are numerous sources of information, all with varying levels of detail, presentation styles, and advice.

CHAPTER 2

Why is Gender Diversity Important?

2.1 INTRODUCTION

Before we jump into what you can do to improve gender diversity, we begin by explaining why gender diversity is an important goal.

There are several arguments we have heard made against actively working toward gender diversity. "We live in a country with abundant opportunities, so why are we trying to make females choose to be in a career they do not want? Shouldn't that choice be more important than whatever desire we have for those bodies to be in computer science?" and "As much as some want to believe females and males are exactly the same, there are biological differences that, for whatever reason, lead to different strengths in females and males. Since males perform statistically better on spatial tests, perhaps computer science is better suited to males."

These arguments against gender diversity assume two things. First, females and males are treated equally. Second, since they are treated equally, there must be innate differences between females and males that cause females to either not be interested in computer science or not be equally capable of succeeding in computer science. If it is lack of interest, then who are we to try to convince females who do not want to be computer scientists that they should be? If it is difference in capability, then it would be counterproductive to do so, requiring the lowering of academic standards to achieve some artificial norm in gender numbers.

Unfortunately, there is ample evidence to show that, despite gains in gender equality in the past 100 years, females and males are treated differently by parents, teachers, and caretakers *from infancy*. While there is little difference in *achievement* between females and males in the math and sciences in K-12 schools (NCES, 2000), there is a large difference in career choices and the perception of others as to their suitedness for top-tier academic success.

In addition, while there clearly are biological differences between males and females, the evidence is very dubious as to whether any of those differences contribute to males' and females' success in computer science. In fact, there is as much evidence to suggest the differences make females *more valuable* employees than there is to suggest they are less valuable employees (which is why companies recruit under-represented minorities so heavily—it's economics, not altruism. See Sections 2.4.2 and 2.4.4). In addition, the overlap between males and females is large enough that even if there were statistically significant differences in the average female and the average male, it would be unconscionable to apply those average arguments to any individual female or male.

Finally, the gender gap itself presents an extra barrier: the lack of females feeds the belief that females are somehow not as talented at computer science, which in turn causes females to do worse in computer science through an effect called stereotype threat. We explain what stereotype threat is and why believing that females (and other minorities) should be represented in every major is important for their success.

Let me just remind you of the limitations of these studies, which we mentioned in the Introduction. First, no statistical study represents everyone, because there is a great deal of variation within the group. When comparing males and females, the overlap is very great. Second, the successful females in computer science were successful for a reason, making them less likely to identify with these studies than the average female. In the future, we hope fewer and fewer females will identify with these studies, because that will signal that the United States is becoming more equitable in terms of the upbringing of male and females.

2.2   COMPUTER SCIENCE–THE WORST IN STEM

The media has published such positive articles about how females have caught up in academia that one might believe parity has been achieved. The number of females receiving bachelor's degrees rose to over 50% for the first time in 2011, National Center for Education Statistics. Sure, computer science appears to be slightly behind, but one might think that, over time, progress is being achieved. It is only a matter of time.

Sadly, that is not the case for computer science. As you can see from Figure 2.1, females have been making fairly steady gains in most STEM fields. The fields with the fewest women have historically been physics and engineering, but they, too, have been steadily increasing to around 20%. Computer science, though, is an outlier. In the early 80s, computer science did very well. In fact, almost a third of

Percent Women Bachelor's Degrees, Selected Fields
1966-2008

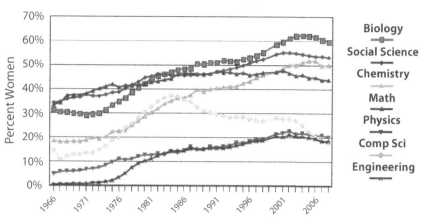

Figure 2.1: Percent Female Bachelor's Degrees, 1966-2008. This shows that, even as other STEM fields have made progress, Computer Science has shown a decline in female representation in Bachelor's degrees since the early 1980's. Source: National Science Foundation S&E Degrees.

bachelor's degrees went to females in the early 80s. Since then, though, females have chosen computer science in *decreasing percentages*. It was not just the dot-bomb (dot-com bust)—*female representation in computer science has decreased for almost 20 years*. Despite the increased effort and funding of gender diversity initiatives, females continue to flee the field.

In a more positive development, Figure 2.2 shows that universities have been successful in gradually increasing the number of female faculty members. Not surprisingly, the percentage of junior faculty has been higher than more senior faculty. We hope these trends continue. It will be difficult, however, if the percentage of females keeps decreasing at the lower levels.

As a result, we feel the most important portion of this guide is the information aimed at affecting the most females at the undergraduate level—in the classroom. However, in order to continue the positive progress at the graduate and faculty levels, we have included tips and background for undergraduate research and graduate research, faculty, and institutional-level changes.

Further reading:

CRA Taulbee Trends: Female Students and Faculty

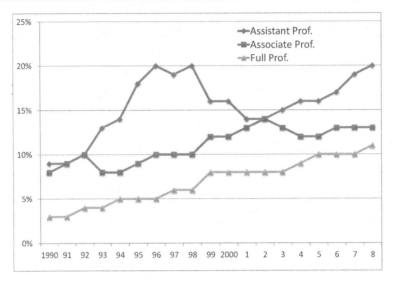

Figure 2.2: Percent Female Computer Science Faculty, as reported by PhD-granting institutions responding to the Taulbee Survey. There has been incremental but steady progress at all levels, from Assistant Professor through Full Professor. Source: Taulbee Surveys.

http://archive.cra.org/info/taulbee/women.html

2.3   THE MYTH OF CAREER CHOICE

The rights of females to equal economic access has been long fought, from Elizabeth Blackwell getting the first medical degree in 1849 to females getting the right to vote in America in 1920. Harvard Law School did not admit females until 1950. In the 1960s and 70s, teacher, nurse, and secretary were often the only careers females believed were open to them. Some now believe that because universities, medical schools, law schools, etc., no longer discriminate on the basis of gender, females are treated equally and have the same opportunities as males.

While the past 100 years has seen tremendous progress, and overt, intentional discrimination is not often an issue, girls and boys still grow up with very different roles, treated differently from infancy by parents and teachers, having vastly different expectations in many religions, and shown different sets of acceptable toys within toy stores. These differences have a profound influence on females' career paths.

The focus of this section is to show that females and males display similar levels of *capability* and *achievement* in math and science, yet make very different career choices. In order to show why, we focus on three points. First, females and males are *expected* to have different interests. These expectations, and the toys they receive, provide different *experiences*, providing different areas of *confidence*. Second, parents provide praise to daughters and sons in different ways, which lead to differences in willingness to take on challenges. Third, their levels of self-esteem are very different in high school and early college, the point at which career choices are made.

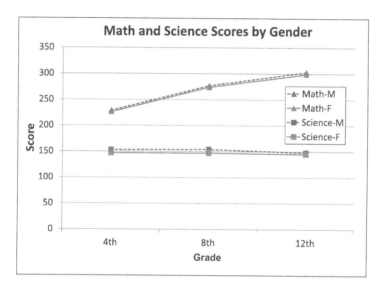

Figure 2.3: Math and science scores for females and males on achievement tests in 4th, 8th, and 12th grade. There is virtually no difference in scores at any of these levels. Source: National Center for Educational Statistics, 2000.

2.3.1 MATH AND SCIENCE ACHIEVEMENT

There are no required computer science courses at any point in education, so we cannot do a broad comparison of female and male potential in computer science. Math and science are the closest fields, so while not perfect indicators, certainly performance in math and science gives insight as to whether there is a predicted gender difference to expect in computer science.

Figure 2.3 shows performance of American students on standardized math and science tests broken down by gender for 4th, 8th, and 12th grade (NCES, 2000). As you can see, there is no statistical difference in performance in math, and only a slight difference in the performance in science. Achievement does not at all explain the gender gap in computer science and engineering.

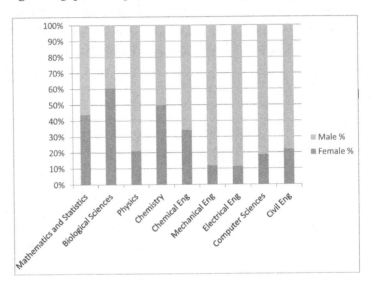

Figure 2.4: The gender balance of bachelor's degrees earned in different STEM fields in 2007. Females are more likely to major in classes they took in high school, leading one to believe *confidence*, not *aptitude* or *interest* is a determining factor. Source: National Science Foundation, Division of Science Resources Statistics, 2009.

If we look more closely at the STEM fields, we see that not all STEM fields suffer a large gender imbalance. Figure 2.4 shows the gender balance of BS degrees in STEM fields. Mathematics, chemistry, and biological sciences have a good gender balance, whereas physics, computer science, and all engineering fields have a poor gender balance. We see that females are much more likely to major in a field that exactly matches a high school requirement than not. This is especially highlighted by the difference between chemistry and chemical engineering. They are both built on the same area, but have quite different representation.

Why is there such a difference between these areas? One aspect is interest, because biological sciences are tied to the medical field, and females have been shown to choose fields that have a large societal impact. Although computer scientists are well aware that CS has a major societal impact (e-mail, Facebook, Google), computer scientists are often portrayed by the media as tech support (annoyingly condescending to everyone else) or hackers trying to use technology for crime. Very seldom are computer scientists using their skills for the greater good (Angela in the crime series *Bones* being a delightful counterexample). Differences in motivation do not explain the gap between chemistry and chemical engineering, nor the gap in general between mathematics and chemistry vs. the engineering and computer science fields.

We propose that this is a gap of confidence and risk-taking. The combination of high school course offerings and college entrance requirements compel students to take mathematics and choose at least two courses in biology, chemistry, and physics. Despite any differences in upbringing and influences about what females *should* do, females receive a somewhat objective, quantitative measure of their achievement in those courses, and they see that they are just as capable as their male peers. Their confidence increases in those fields, and they are more likely to major in them. Without required courses in engineering and computer science, females do not have that opportunity. Thus, majoring in those areas is a risk and requires confidence.

2.3.2   THE RISK-TAKING GAP

Risk-taking might be somewhat innate, but caretakers have great influence on a child's willingness to take risks and challenge themselves. Carol Dweck of Stanford has pioneered recent work on the effect praise has on behavior. She has found in several studies that praising ability rather than effort causes children to attempt easier problems in order to reinforce their identity as smart rather than attempting harder problems to show how hard they try.

She has a new long-term study (Gunderson et al., 2013)—the first to observe and measure effects in normal family interactions rather than a laboratory experiment. Her team observed the ratio between praise of personal traits (i.e., smart) vs. effort in 1-3 year olds and measured the child's willingness to attempt challenging problems five years later. They found two results. First, the more that effort was praised, the more likely students would take on challenging problems and take risks five years later. Second, males were more often praised for their effort, and females were more often praised for their personal traits. Thus, males are being nurtured to take more risks and tackle challenges from before they can talk. This effect is shown in Figure 2.5.

In adulthood, women have been found to be more risk averse to men in 150 studies, from making financial decisions (Jianakoplos and Bernasek, 1998) to playing chess (Gerdes and Gransmark, 2010). A meta-analysis

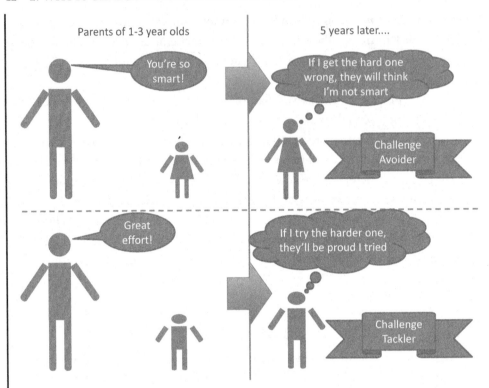

Figure 2.5: The type of praise given to children has lasting effects 5 years later. Females are more often given praise about personal traits, leading them to reinforce that through low-risk endeavors with high-probability success. Males are more often given praise about effort, leading them to reinforce that through taking on challenges. Source: Gunderson et al. (2013).

(Byrns, Miller, and Schafer, 1999) found that studies overwhelming show a real gap between males and females. The gap is situational, more pronounced in areas such as intellectual risk taking and physical skills, and less pronounced in others such as smoking. In addition, the gender gap is shrinking over time, showing that it is neither an innate trait, nor is the gap unfixable.

Further reading:

Babies whose efforts are praised become more motivated kids, say Stanford researchers

`http://news.stanford.edu/news/2013/february/talking-to-baby-021213.html`

2.3.3   THE EXPERIENCE GAP

These same children are receiving much different toys to play with. Toy stores often have toys sorted by gender, with blue sections and pink sections. Sadly, an attempt by Hamleys, London's best-known toy store, to introduce gender-neutral signs and maps, was abandoned after eight months (Bennhold, 2012). The boys' section has building toys such as LEGO's,[1] tinker toys, tool tables, and marble ramps. The girls' section contains kitchens, dollhouses, dolls, and stuffed animals. Thus begins both the socialization of what girls and boys *should* do and the practice and confidence-building influencing what girls and boys *can* and *will* do.

Nobel prize winner Steven Chu (1997, Physics) says in his autobiography: "The years of experience building things taught me skills that were directly applicable to the construction of the pendulum. Ironically, twenty five years later, I was to develop a refined version of this measurement using laser cooled atoms in an atomic fountain interferometer (Chu, 1997)."

By providing boys years of play with building toys, they increase their skills and confidence in building things, which is what we are taught is the basis for engineering. Computer science majors are very often housed in the College of Engineering. In addition, computer programming has a long history of the "tinkering" approach, very much akin to taking apart watches and radios to learn them. This entire philosophy of learning is much more similar to tactile building than it is to social play. Thus, even though there are blue computers and pink computers for girls to *use*, the programming of computers continues to be presented in a way that resonates with male toys.

Perhaps without this experience and confidence, girls are less likely to choose this as a major, either consciously fearing they are already years behind or merely lacking the positive experiences that provide self-efficacy in this area.

In fact, a study from Carnegie Melon University (CMU) found that most females in computer science were one of three groups: is the oldest in the family, has no brothers, or is there for economic reasons (Margolis and Fisher, 2003). For the first two cases, it is likely that the girl had more time with her father, experiencing the style of play boys typically get.

Further reading:

Unlocking the Clubhouse, Margolis and Fisher
`http://dl.acm.org/citation.cfm?id=543836`

---

[1]LEGOs introduced pink LEGOs, lowering the barrier to people buying LEGOs for girls. With LEGO Friends, the themes have changed, and perhaps this will narrow the gap for future girls.

Toys Start the Gender Equality Rift

http://www.nytimes.com/2012/08/01/world/europe/01iht-letter01.html

2.3.4   THE CONFIDENCE GAP

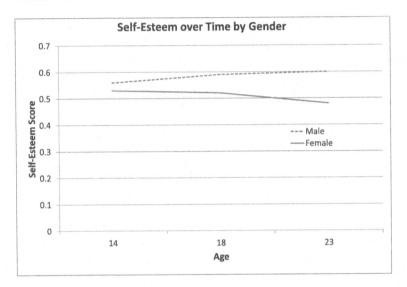

Figure 2.6: Self-esteem of students in middle school through college years, showing that females have much lower self-esteem than males when making major career choices. Source: Block and Robins (1993).

By the time students enter elementary school, girls are more risk averse, and boys have more experience with engineering- and computer-related toys. We do not mean to imply these are the only differences, only the ones on which we focused. These two (and possibly other) factors reinforce themselves, building a cycle in which boys further build their skills and confidence because they challenge themselves more.

Treatment in school reinforces these differences. Teachers ask boys more questions (French and French, 1984), and the type of question is more open-ended and challenging than those asked of girls (Swann and Graddol, 1998). Thus, females have less practice with open-ended problems, leading to less confidence in these types of tasks.

As students hit puberty, confidence reaches an all-time low. Both females and males report very low levels of self-esteem (Figure 2.6). In high school, though, the differences become striking again. The self-esteem of males begins to recover (Block and Robins, 1993), while female self-esteem continues to fall. It is at this time, when the gap in self-esteem is the largest, that males and females must make career choices.

The effect of self-esteem does not just affect overall confidence. It affects students' perceptions of what they are good at, which in turn affects in what careers they think they would succeed. These also affect perseverence within the major, which we will address in Chapter 3.

So, despite equal aptitude in math and science, females have gaps in risk-taking, experience, and confidence in computer science and engineering fields. Females tend to choose majors in which they can make a positive impact on society (education, medical care), fields in which they stereotypically do well (writing-based humanities), or ones they have already received quantitative feedback that they can do well in (mathematics, biology, and chemistry). Unless we change the messages we are sending, or require all students to take a course in order to provide that positive feedback, we cannot expect females to choose engineering and computer science, regardless of ability.

2.4   WEIGHING GENDER DIFFERENCES

Despite females now achieving at the same level of males in math and science (as shown in Figure 2.3), some people argue that there might be a biological reason that females are not as good at engineering, and the gender gap is justified in this way. The argument is that there are clearly physical differences between males and females, so there are probably cognitive differences, as well. These might make females and males better suited to different careers.

If this were true, then businesses would not be striving so hard to increase the diversity of their work forces. Businesses are not doing this out of an altruistic sense of fairness—they care most about their bottom line. Businesses want a diverse workforce for three major reasons. First, they want their products to appeal to a diverse population, so it is important that the designers reflect the consumers. Second, regardless of what the product is, a diverse team has more variation in ideas, leading to a better solution. Finally, in teamwork, it turns out that there is some evidence that having more females leads to more productive teams. The benefits of having female engineers greatly outweighs the evidence of any biological disadvantage females may have.

It is impossible to determine, one way or another, whether females or males are better suited biologically to computer science. Teasing out the distinction between genetic/biological vs. environmental factors is extremely difficult. Instead, my goal is to present evidence that, at best, shows that females are as much at an advantage as disadvantage, and at worst, exposes reasonable doubt. We believe reasonable doubt is all that is necessary because the injustices caused by believing (and acting on a belief of) innate differences are far too damaging to be justifiable in the presence of reasonable doubt.

2.4.1   DISADVANTAGE – SPATIAL ABILITY

This subject is controversial, and scientists have a poor historical track record in their attempts to "prove" superiority of one group of people over another with respect to cognitive tests. *The Mismeasure of Man* (Gould, 1981) presents historical attempts to prove that Africans are less intelligent than Caucasians because of smaller cranial capacity. First, the scientists' methods were compromised by their beliefs, leading them to (unintentionally?) pack the material tighter in Caucasians' skulls. Second, cranial capacity is not the determining factor in intelligence. He further argues that even with various intelligence tests, the ties between those tests and success in certain fields are not as closely linked as scientists would like to believe.

We find ourselves in a similar position with respect to gender and computer science. Males tend to do better on spatial tests than females. Females tend to do better on "social intelligence" tests than males. The two questions, then, are whether the performance is biological rather than learned, and whether people with higher spatial rotation ability are more suited to computer science than those with higher social intelligence. Finally, we must always recognize that the overlap between genders might be larger than the differences. So even in the presence of differences in the average person, a large gender gap may not be warranted.

When making biological arguments, there are generally two schools of thought. The first is evolutionary psychology, arguing that evolution favored certain traits in females and males based on their gender roles in prehistoric times (i.e., males are better at judging faraway targets for hunting, and females are good at short-distance focusing for gathering (Stancey and Turner, 2009). The second is that the chemical differences in the bodies affect the skill (i.e., testosterone makes males stronger and more aggressive).

Let's begin by exploring the evolutionary psychology argument. The field itself, from its inception, has been controversial. Controversy about the conclusions drawn using this paradigm aside, in order for this argument to be valid, there must be a compelling reason that, based on the roles of females and males in the hunter-gatherer era, males needed spatial rotation ability more than females. At the 2012 NCWIT Summit, Dr. Nora Newcombe argued eloquently against this position (Newcombe, 2012).

Is spatial rotation ability more important for hunting than for gathering? Hunters must memorize the locations and migratory patterns of animals, coordinate attacks, and fashion weapons. Gatherers must memorize the locations of water and plants, know which plants are ripe at what time, and plan and execute gathering routes appropriate to which plants are currently ripe. While we could bicker about the details required for each task, and which requires slightly more spatial ability than the other, we think it is clear that gathering over large distances on a daily basis does not require significantly less spatial ability than hunting large game on a less-than-daily basis. Therefore, evolutionary psychology does not explain the difference in spatial ability.

The second argument is that the testosterone males have improves their spatial ability. There is variation within males and females as to how much testosterone they have. For this to be true, larger amounts of testosterone should correlate with higher spatial ability. Studies have shown mixed results. Some compare the same males as their testosterone fluctuates during the day (with highest testosterone in the morning). These show a correlation between higher testosterone in the morning and better performance on spatial rotation tests (Silverman et al., 1999, Moffat and Hampson, 1996). It is unclear whether this is caused by testosterone or general fatigue. When comparing different males, they have found an inverse correlation between testosterone levels and spatial ability (Moffat and Hampson, 1996). In females, spatial ability has been shown to reduce with an increase of estrogen (Elizabeth Hampson, 1990), when comparing the same females during different parts of their menstrual cycle. At this point, scientists do not have a unified theory. The fact that when comparing different males, lower testosterone correlates with higher spatial ability means that a large number of females with high testosterone (for a female) could be equally advantaged as males with high testosterone.

Given how inconclusive these studies are, one cannot discount the differential upbringing females and males have. In particular, the toys in the "boys section" of stores include LEGOs, erector sets, and the like. As quoted earlier, Steven Chu,

Table 2.1: Examples of design errors due to lack of female engineers. Source: see text below

Product	Shortfall	Effect
Cars	Lack of visibility for short drivers	More accidents Negative driving stereotypes
Voice Recognition	Did not recognize female voices	Unusable by females
Airbags	Not calibrated for short drivers, passengers	Unnecessary deaths of short women and children

1997 Physics Nobel Prize winner, credits his later success on his early childhood building experiences (Chu, 1997). By providing boys years of play with building toys, they increase their spatial skills. Girls, on the other hand, are steered toward nurturing roles, providing practice with different skills.

Finally, regardless of the reason for the difference in spatial ability, we would still need to believe that spatial ability is correlated to success in computer science for this to be a reason females are ill-suited to the field. In addition, the difference in representation should be largely similar to the difference in spatial ability. The gap in spatial ability, though, is much smaller than the gap in representation. In fact, with a little training, females are able to close this gap tremendously (Newcombe, 2012). There are more than enough jobs available for the range of spatial ability achievable when females and males receive training in spatial tasks.

Further resources:

Nora Newcombe's 2012 NCWIT Summit Talk

`http://vimeo.com/channels/372194/45873134`

2.4.2 ADVANTAGE – DIVERSITY OF EXPERIENCES

The biological disadvantage of being female has certainly not been proven. There is actually much more evidence that being a female is an advantage *because of their minority status* as well as differing strengths. Let me be clear—we are not advocating the view that females are superior to males, any more than that males are superier to females. We are merely presenting this evidence to show that there is at least as much evidence that products would benefit from the inclusion of more females (but not necessarily more than a 50/50 split) as that products benefit from mostly male designers.

Engineers design a plethora of products that target the general population. In the design of these products, there are fundamental physical differences between females and males that should not be overlooked (i.e., size, strength, voice) as well as tastes and preferences, that affect the suitability for females. Engineers design critical equipment such as medical devices, wheelchairs, automobiles, and air bags, for which these physical differences are important. Some serious shortfalls are shown in Table 2.1. For example, air bags were initially designed for adult male bodies, leading to preventable deaths of women and children (Margolis and Fisher, 2003). In a less critical example, the Volvo YCC was designed by women to showcase the ergonomic perspective of a female driver. Among the features were headrests with space for a pony tail, the front lowered and fenders put in view for short drivers to allow the driver to see where the four corners of the car are, and pushing a button opened the nearest door to allow someone juggling bags to easily place things into the car (Volvo, 2004). Finally, early voice recognition systems were calibrated to low male voices, making children and women's voices unrecognizable to the system (Margolis and Fisher, 2003).

These are just a few examples of how including or excluding females from design teams can have profound effects on the usefulness of the products to 50% of the population. If companies want to design products for the general population, it is critical that females (and other minorities) be represented.

Further Reading:

Unlocking the Clubhouse, Margolis and Fisher
http://dl.acm.org/citation.cfm?id=543836
Volvo YCC details
http://www.howstuffworks.com/volvo-concept.htm

2.4.3   ADVANTAGE – DIVERSITY OF IDEAS

Theoretically, one could design a product for diverse populations, not by having the engineers be diverse, but by having a diverse set of customer consultants that the design team could interview for their feedback on design ideas. But designing for the diverse customer is just one advantage of having a diverse team. Diversity of ideas also leads to better solutions.

Why do we work in design teams rather than having a single person design a product? Because businesses for years have recognized that, in most cases, the product will be better if you have multiple people proposing ideas. The goal is not just to increase the total number of ideas, but to increase the differences between those ideas. It is not just the intelligence of the individuals that matters, but the

Table 2.2: Just a few of the creative ideas about rental behavior. Source: van Buskirk (2009)

Netflix contest diverse rental pattern ideas
Weekend rental behavior different from weekday rental behavior
Abrupt changes occur when people start, end relationships
People rating movies they saw long ago rate differently than ones just watched
People who rate a slew of movies at once are rating ones not watched recently

diversity of the ideas. In fact, many ideas that are more different from each other can be more useful than a few excellent ideas.

This concept is not just theoretical. An amusing application of this is a recent Netflix Prize. The rough details are the following: Netflix has an annual contest to predict, based on prior ratings, what people will like. In order to win the contest, the submitted software must be 10% better than Netflix's existing software at predicting what a user will want to watch. There was a million-dollar prize for the winning team.

Different groups were competing individually, and while they initially made great strides, the improvements were stagnating. Then some teams started working together. Interestingly, the teams that contributed most to general improvements were those with algorithms that were most different from the others (van Buskirk, 2009), shown in Table 2.2.

The key to success for these algorithms is that the people in the group need to have sufficiently different ideas. Different ideas come from different upbringings, experiences, etc. Those with very similar experiences tend to have very similar ideas, which might make for a fun, positive environment, but ultimately results in a weaker outcome. It is important that we learn to celebrate the diversity of backgrounds and ideas.

Further Reading:

How the Netflix Prize was Won

http://www.wired.com/business/2009/09/how-the-netflix-prize-was-won/

Scott Page: The Difference: How the Power of Diversity Creates Better Groups, Firms, Schools, and Societies
    `http://ww2.ncwit.org/pdf/NCWITSummit_ScottPageSlides.pdf`

2.4.4   ADVANTAGE – SOCIAL INTELLIGENCE

Not only is there a general argument that products are stronger with diversity of ideas, but there are research and real-world examples that show positive effects with high representation of females.

In the Lab

An intriguing study about *collective intelligence* (Wolley et al., 2010), as opposed to *individual intelligence*, explored what individual characteristics led to better group results. There has been a lot of research on individual intelligence and how performance on different intelligence tests correlates with performance solving problems. The question they wanted to answer, though, was whether performance on different intelligence tests correlated to performance solving problems in a group. Participants were found by advertising on Craigslist, and they were randomly placed into groups based on when they could make appointments to participate in the study. Participants were given individual intelligence tests, then they were asked to solve different problems as a group. The groups varied from three to five members. They then looked at a variety of factors and looked for a statistical correlation between those factors and end performance.

They found three factors that correlated with success. The first was gender. Groups with at least 40% females performed better. In addition, groups with more individuals that have high performance on the social intelligence test did well. Finally, groups that took turns speaking performed the best. Two of the three factors were correlated—females tended to have the higher social intelligence scores. Other individual intelligence tests did not have a large effect on the group's performance. We can only go so far with this research—in companies, people work together for many years, so a group project solved by strangers is not necessarily the closest environment to a business environment.

Further Reading:

Defend Your Research: What Makes a Team Smarter? More Women, Woolley
    `http://hbr.org/2011/06/defend-your-research-what-makes-a-team-smarter-more-women/`

In the Real World

It turns out, though, that there is evidence even in the real world that gender can influence performance. For example, Erhardt et al. (2003) found that a percentage of women and minorities on boards of directors for 127 large US companies was positively associated with financial indicators of firm performance (return on asset and investment).

This occurs for one of two reasons—the females on the board make important contributions to the company, or a company whose culture is such that females are on the board is also a company with better female representation at all levels in the company. Either way, this implies that more female participation *at the upper levels of businesses* leads to financial success.

A second example deals with Wall Street superstars who switch companies. Groysberg (2010) performed a detailed analysis of the careers of more than a thousand star analysts on Wall Street. He found that performance is about more than the individual, because those who were individually lured away to another company suffered a drop in performance for an average of five years. Those whose whole units were sold to/acquired by another company showed no change in performance. When those who were individually lured away were divided by gender, it turned out that females took only an average of three years to recover prior stellar performance. This led to two conclusions. First, Wall Street performance is highly dependent on the research departments that support those stars. Second, females are better at managing and/or training those groups in a short amount of time.

Further reading:

COLLECTIVE INTELLIGENCE: "The Influence of Women on the Collective Intelligence of Human Groups," Dr. Chris Chabris

`http://ww2.ncwit.org/pdf/CollectiveIntelligent-NCWIT-20120523a.pptx.pdf`

2.5   CONCLUSIONS

In summary, we have provided evidence that:

Females may avoid computer science and engineering because of societal influences, not capabilities

- Females and males have nearly identical achievement in math and science through K-12

- Differences in praise is correlated to differences in the willingness to take risks and attempt challenging tasks

- Differences in toys may lead to differences in experience in physics, engineering, and computer science

- Differences in treatment may lead to differences in confidence

  There is little to mixed evidence that males have any biological advantage based on spatial ability

  - Evolutionary psychology does not provide evidence for gender differences in spatial rotation ability

- Biological studies are mixed on whether biological factors explain the differences in spatial rotation ability

- Domain knowledge outweighs advantage of spatial rotation ability

- No studies have shown the difference in spatial rotation ability makes a difference in succeeding in computer science

- No studies have shown that the difference in spatial rotation ability is larger than the gender gap in computer science

  There is some evidence that females have advantages by being a minority and because of social intelligence

- Product teams with diverse members produce stronger products that are suited to more of the population

- Females tend to have higher social intelligence

- In a laboratory environment, females work in teams better than males

- In the real world, some examples show that females lead teams better than males

  Scientific data shows that females are at least as valuable as males for computer science companies and interdisciplinary, competitive research, yet have significant hinderances to choosing computer science

  Since neither side can be proven, why are we not satisfied to allow us to just agree to disagree?

It is harmful for females if you behave as though you believe any of these arguments or if you perpetuate the societal stereotypes that deter females from joining computer science. You do not need to implement the majority of the suggestions. There is a wide spectrum, from being part of the problem, to being neutral, to being part of the solution. If you believe females are less capable and act on those beliefs, then *you are part of the problem.* At the very least, professors have a moral obligation to be neutral and give all students fair evaluations.

If you believe females have free choice of major, then you may not look and see when a female is making a choice based on low self-esteem instead of what she is capable or would enjoy. It is important that faculty members actively counteract the social forces that cause females to question their ability in computer science. We will go into more detail about this in the advising and classroom sections.

If you believe that females biologically are not as capable of succeeding in computer science, then you might be more likely to advise a female to leave the field or give subtle indicators that she will not succeed. Even if it were true, the differences within females and males is much larger than the differences between them, so this small statistical difference could not be used on any individual.

In addition, knowing the economic arguments and conveying those to students can allow underrepresented students to see the value in themselves—something they bring to the table that no one else can. This increases their feeling of worth, counteracting some of the low self-esteem.

The lack of females shapes young males' impressions that females' talents lie in staff roles, not technical roles. Male undergraduates see that the only females in the department are staff, reinforcing or creating an impression that females are not suited to computer science. When those males become managers, they may marginalize females, giving them support roles (as has been found).

Finally, the lack of females in computer science makes it harder for those there. Not just because it is lonely, not just because it is harder to find study partners. It perpetuates a negative stereotype, and the negative stereotype in itself can negatively affect performance. In a nutshell, stereotype threat is anxiety brought on by being reminded that one is not supposed to be as good in computer science. This is especially troubling because the *belief* that one's group does not do as well overall *causes* individuals in the group to do worse.

Further Reading:

Why So Few?

http://www.aauw.org/files/2013/02/Why-So-Few-Women-in-Science-Technology-Engineering-and-Mathematics.pdf

CHAPTER 3

Obstacles to Gender Balance

The obstacles to gender balance today are much harder to see than a generation or two ago. When universities would deny entrance because their "female quota" had already been satisfied, or bosses would not hire females because they thought they would just quit to have a family, it was clear that the environment was different for females and males.

In Section 2.3, we presented three gaps caused by the different upbringings experienced by females and males—the risk-taking gap, experience gap, and confidence gap—and how they make females less likely to *choose* computer science as a major. In this chapter, we address how these same factors negatively affect female students *pursuing* computer science, both in college and in their careers. In addition, the environment itself is very different in college than it was in K-12, providing an external obstacle for females.

If you would like to jump right to the tips, we invite you to do so. The tips refer back to these sections, so you might prefer to look at *what* we suggest before *why* we suggest it.

3.1   INTERNAL OBSTACLES

We will first discuss what could be described as internal obstacles. After 18 years of being treated differently, females have been trained to feel and act differently in the same circumstances as males. These external influences have now been internalized, and they become internal obstacles. Being able to recognize those obstacles does not just help solve the gender gap—it also helps males who have similar internal obstacles. The key is that recognizing and handling these obstacles allows many students, male and female, to be more likely to succeed in computer science who could not for reasons completely unrelated to their technical potential.

3.1.1   RISK-TAKING OBSTACLE

As described in Section 2.3.2, women are less likely to take risks than men. In K-12, this is a positive trait, because females are more likely to start homework early, complete it thoroughly and on time, etc. As they progress to higher education, more emphasis is placed on thinking outside of the box and independent work, both of which benefit from being willing to take risks. The NSF (National Science Foundation) puts a high premium on transformative research, as opposed to incremental research. Those unwilling to take risks are less likely to excel at the upper echelons of academia, reducing females' chances of success.

A lack of confidence has some major effects that professors might not be aware of. Students might choose easier projects when given the option (in order to guarantee success). These behaviors lead professors to believe the students are less intelligent, knowledgeable, creative, or motivated because of their conservative projects.

The good news is that the gap is shrinking (Byrns, Miller, and Schafer, 1999). This means the behavior is learned, so it can be unlearned. By emphasizing failure as a normal, even necessary part of the research process, we can encourage those who are risk averse to take more risks.

Further Reading:

Why Women Don't Take Risks With Their Money, Helaine Olen—this is about investing, not education, but it shows how the same data can be interpreted as an innate or learned trait. There are compelling reasons why females act differently than men in a society that is not yet equal.

http://www.theatlantic.com/sexes/archive/12/11/why-women-dont-take-risks-with-their-money/265224/

3.1.2   CONFIDENCE OBSTACLE

Figure 2.6 showed the large gap in confidence between females and males in high school and college, influencing females' decisions not to major in computer science and engineering. One might conclude, then, that females who pursue computer science and/or engineering are as confident as their male peers.

One might think that only the confident females chose computer science, so the ones in the field are not different from the males. There are signs that even the females who choose computer science are less confident than male counterparts.

Upon joining a computer science major, those females will find themselves a minority, perhaps for the first time in their lives. In addition, males are more likely to have programming experience entering college (Margolis and Fisher, 2003), magnifying this sense of isolation for females. Finally, females transfer out of the major, *at higher GPAs than males*, citing failure as the major reason (Patterson and Trasti, 2004, Margolis and Fisher, 2003). Therefore, it is likely that the average female you encounter will have less confidence in her place in the major and performance in the major.

A lack of confidence has some major effects that professors might not be aware of. Students who are less confident tend to be more risk-averse, choosing easier projects when given the option (in order to guarantee success), and they ask questions to get confirmation even when they actually know the answer (because they doubt themselves). *These behaviors lead professors to believe the students are less intelligent or knowledgeable than they really are.*

Thus, the challenges are great. There are several things professors can do to distinguish between lack of confidence and lack of knowledge as well as structure communication and assignments in order to reduce the effect confidence has on behavior (See Chapter 4). These techniques are not only useful for females, but for any student who lacks confidence in his or her abilities.

3.1.3 COMMUNICATION OBSTACLE

Differences in communication patterns taught from a very young age have a negative effect on females both in group projects and when interacting with faculty. Females more often express submissive emotions, ones that do not negatively affect others, such as sadness and anxiety. Males, on the other hand, display strong, combative emotion when necessary (Zahn-Waxler, Cole, and Barrett, 1991, Brody and Hall, 2000). This plays out to females' disadvantage in many ways.

Tannen observed a number of differences in corporations when observing mixed-gender meetings (Tannen, 1994). Differences included men phrasing statements in ways that claimed attention and credit for their contributions. Females were less likely to strongly assert their opinions, prefacing statements with a disclaimer even when they are right, as shown in Figure 3.1. This leads to lower likelihood that females' ideas will be pursued, reducing their contributions in the group, and therefore the perceptions of their value.

Figure 3.1: Illustration showing how different communication patterns may lead to inaccurate perceptions of female knowledge.

In addition, when discussing ideas, men were more likely to do so in a challenging, somewhat adversarial style. They believed that this style led to the best resulting ideas as people thought of the advantages and disadvantages of their ideas. Women, on the other hand, were more likely to take these adversarial statements as indications of the weakness of their ideas or personal attacks. This may lead them to be reticent to pose ideas if they are criticized so strongly.

Females are less likely to highlight their accomplishments. Thus, males' accomplishments are more known to the group, and superiors get the impression that the males are contributing more and succeeding more.

Finally, females are more likely to express self-doubt. These seemingly self-aware expressions may color others' views of them, convincing them these doubts are actually true, as shown in Figure 3.2. It is challenging to distinguish between the words that we hear and observable performance. What may sound to a professor like very self-aware knowledge about the student's limitations might actually be a communication pattern.

These problems can be lessened by providing communication instruction for all students to make sure that less assertive students assert themselves and more

Figure 3.2: Illustration showing how different communication patterns may lead to inaccurate perceptions of female promise.

assertive students wait to hear everyone out. Groups and departments can highlight all accomplishments by having an individual in charge of gathering awards and highlights. It can be advertised as important for the visibility of the department to find out about these successes. Then these are circulated within the department as well as beyond so everyone is aware of successes. Finally, professors need to be trained to recognize and not be colored by expressions of self-doubt. Instead, they should look to objective measures of achievement. Otherwise, these will lead to fewer research opportunities and weaker letters of recommendation.

Further Reading:

Talking From 9 to 5: Women and Men at Work, Deborah Tannen—A leading researcher in gender communication combines anecdotes and research to illustrate the differences in gender communication patterns. This is useful for male or female managers to see past patterns to the true conent, as well as females to interpret what others really mean and adjust their own patterns so they are not misunderstood.

Lean In, Sheryl Sandberg—In this book, Sandberg chooses to focus on what women can control, which is their own actions. While some believe this is placing the blame on women, instead it is intended to empower women to change their communication patterns (among other things) in order to succeed in the workplace.

3.2   EXTERNAL OBSTACLES

External obstacles are factors that are present in computer science college educations or workplaces that were not present in K-12. These are new challenges that provide an extra hurdle for females.

In K-12, 84% of teachers are female (C. Emily Feistritzer, 2011). In computer science, this ratio is inverted—82% of teachers are male (Taulbee, 2010).[1] In addition, most K-12 classrooms are fairly even in the number of males and females. In computer science, however, females constitute 11.7% of CS Bachelor's degrees awarded in 2010-2011. In addition, faculty spend much less time with students (and have many more of them), so they form opinions based on much less information. These external factors and changes—going from equal representation to being a minority, and the gender swap of the evaluators—has a large effect on females.

3.2.1   MALE-DOMINATED EVALUATORS AND PEERS

The communication patterns that females use are not necessarily a disadvantage when working with a random sampling of females. But in a field that was exclusively male, male communication patterns became the norm. Even females who succeed are likely to exhibit male communication patterns, since the ability to communicate this way may have contributed to their success (Tannen, 1994).

Therefore, entering a male-dominated field places females using female communication styles into a situation where the norm is a different style. These differences put them at a disadvantage. The fact that the evaluators are more likely to exhibit and be used to male communication patterns means that females' communications are much more likely to be misinterpreted (as seen in Figures 3.1, 3.2).

Communication and behaviors affect students not only with direct interactions between faculty and students, but also in group work. Students are working in

---

[1]This leads to questions about whether males are at a disadvantage in K-12. This is an important and interesting question but beyond the scope of the book and not relevant to the gender gap in computer science.

groups in undergraduate classes and graduate research groups in a much less structured manner than in industry. Even in industry, though, females suffer, especially when they are in the minority.

Kanter (1977) describes *tokens* as persons who belong to a social category that constitutes less than 15% of the entire group composition. When women are the tokens in a working group, they are more likely to have their mistakes amplified, be socially isolated, and be found in roles that undermine their status. This has been found across a large variety of careers (Mcdonald, Toussaint, and Schweiger, 2004), and only female tokens are at a disadvantage (when men are the minority, they do better than the females in the group).

These group dynamics are caused by differences in communication patterns that are amplified as the ratio of females to males is decreased. Men not only interrupt women more frequently than women interrupt men, but men yield to other men more often than they yield to women. Women, on the other hand, yield to men and women equally (Smith-Lovin and Brody , 1989). In addition, in a brainstorming session, the response to ideas can be a vigorous debate (male style) or conversational refining (female style). One who is used to a conversational style may feel attacked in the debate style, leading them to stop participating. Several studies have suggested (Tannen, 1994, Case, 1994) that this mismatch favors males, making it more likely that their ideas will be pursued. Thus, the more male dominated the group is, the less a women will both be *allowed* to and *desire* to contribute.

This has a profoundly negative effect on women. Being shut out of the brainstorming and decision-making process means that fewer of her ideas will be tested by the group, lowering the likelihood she will get first-author publications, and leading to a lower reputation within the group. This can be mitigated by explicit leadership roles for females (Mcdonald, Toussaint, and Schweiger, 2004), mindful facilitation of group discussion, and explicit communication training.

3.2.2 THE NEW SEXISM: GENDER ROLES OBSTACLE

Many people believe that because *hostile sexism*, for example firing women when they become pregnant or get married, or not hiring them in the first place because they may get pregnant, is rare—equality has been attained. Despite the recognition that women should be judged equally, sexism is alive and well, but it is often unintentional.

Glick and Fiske (1996) defines *benevolent sexism* as a "a set of interrelated attitudes toward women that are sexist in terms of viewing women stereotypically and in restricted roles but that are subjectively positive in feeling tone (for the perceiver) and also tend to elicit behaviors typically categorized as prosocial (e.g., helping) or intimacy-seeking (e.g., self-disclosure)."

These include statements about females or directed to a particular female. Examples would be telling a female computer science student she should take notes because she is so good at it (with no more knowledge about her than her gender), compliments on a female's appearance (hair, clothing, jewelry) after a technical talk, obituaries that, regardless of a woman's accomplishments, list her motherly and wifely accomplishments first, and the number of fans who wrote in their shock at how "hot" the creator of the Facebook page "I F-king Love Science" is. An excellent article about these effects was recently updated because of two striking events: Yvonne Brill's death and obituary and Elise Andrew's revelation that she created I F-king Love Science (Tannenbaum, 2013).

Even when the authors of these comments were told of the inappropriateness of their comments in terms of sexism, they struck back stating they had done nothing wrong.

Unfortunately, statistics indicate these "compliments" may not be benign. Glick and Fiske found that across 15,000 men and women in 19 different countries, endorsement of benevolent sexism was correlated with explicit, hostile attitudes toward women. More troubling, it was also a significant predictor of nationwide gender inequality, even when controlling for hostile sexism. In those countries, women had marginalized roles, less education, and shorter lifespans than men.

In computer science, more likely effects are an effect called backlash—pushing females into more "female" roles (i.e., note-taker, organizer) and punishing females who excel in "male" technical roles. This has been found to occur in groups when females comprise a very small percentage of the group members Kanter (1977), Mcdonald, Toussaint, and Schweiger (2004), as described in Section 3.2.1

When females do take on more male communication patterns, earning leadership positions, asking for more compensation, switching jobs, a backlash can occur. The conventional wisdom is that the gender pay gap is because women are less willing to switch jobs, ask for more compensation, or ask for a promotion. A new report from nonprofit research group Catalyst found this not to be true. Instead, women are more often *denied* these requests (Catalyst, 2011).

Once women break out of jobs more closely fitting traditional gender roles, there becomes a mismatch between what a woman is perceived to be like and conceptions of what she should be like (Heilman et al., 2004). In fact, there are tips on how females can perform the work of males but appear to be performing the work of females (Williams and Dempsey, 2011).

To get hired and promoted in academia, letters of recommendations written by mostly male peers and advisors introduce damaging subjectivity when evaluating a females' potential as top-tier researchers. This can be mitigated by mindful writing and reading of letters of recommendations combined with more emphasis on quantitative measures such as quality and quantity of papers authored (See Chapter 6).

Further reading:

Benevolent Sexism

`http://blogs.scientificamerican.com/psysociety/2013/04/`
`02/benevolent-sexism/`

Self Promote without the Backlash

`http://www.huffingtonpost.com/joan-williams/womens-`
`career-advice_b_1029059.htmlr`

From Marissa Mayer to Sheryl Sandberg: Undeserved Backlash Against Women in Business?

`http://www.openforum.com/articles/an-undeserved-backlash/`

NCWIT: How Does Combating Overt Sexism Affect Women's Retention? Assessments for Identifying Overt Sexism

`https://www.ncwit.org/resources/how-does-combating-overt-`
`sexism-affect-womens-retention-assessments-identifying-overt`

3.2.3 STEREOTYPE THREAT OBSTACLE

Stereotype threat is both an external and interal obstacle. It is caused by both previous and *current external conditions*, but the effect is to change one's *internal actions.*

Stereotype threat refers to *being at risk of confirming, as self-characteristic, a negative stereotype about one's group* (Steele and Aronson, 1995). Just the reminder of the stereotype, and one's inclusion in the group, has been shown to negatively affect one's performance in that activity. In fact, this has been shown in 300 research experiments. In addition, test performance is not the only negative effect.

Table 3.1: Studies showing measurable effects of Stereotype Threat in varying circumstances. Source: Aronson et al. (1999)

Target	Situation
Females	Math tests, Political knowledge, Driving tests, Chess, Computers
African Americans	Mini-Golf
Caucasian Males	Math tests against Asian males, social sensitivity
Latinos	Verbal tests
Elderly	Short-term memory tests

This has been shown for African Americans and Hispanics taking tests (Steele and Aronson, 1995, Schmader and Johns, 2003), females taking math tests (Inzlicht and Ben-Zeev, 2000, Good, Aronson, and Harder, 2008, Spencer, Steele, and Quinn, 1999), and, just to show that anyone can be affected, white males when told they were competing against Asians in math (Aronson et al., 1999). Table 3.1 shows a partial list of experiments that have shown a measurable effect from stereotype threat on specific groups of people in specific situations.

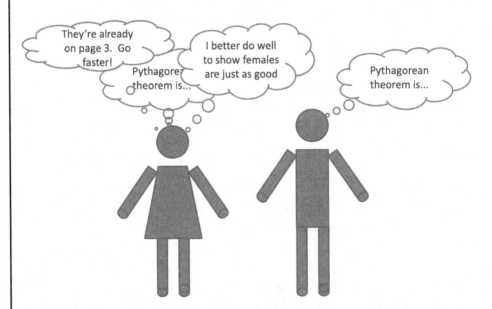

Figure 3.3: Illustration of how anxiety caused by stereotype threat might lower the achievement scores of students on high-stakes tests.

It is not known exactly how stereotype threat lowers performance. Somehow, the awareness that one is a member of a group that stereotypically does poorly in this area interferes with one's performance. This anxiety could cause hormonal changes or distract the student with thoughts during the test, as shown in Figure 3.3.

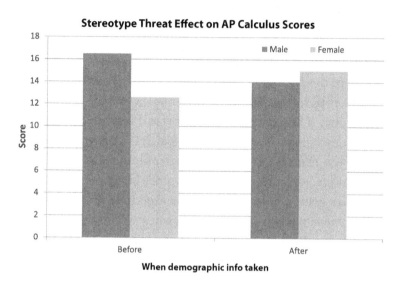

Figure 3.4: Scores on AP Calculus test when demographic info is taken before vs after the test. Source: Stricker and Ward (2004), Danaher and Crandall (2008).

The effect is in no way minor. One study was performed with the AP Calculus test, randomly choosing whether to ask demographic information (including gender) before or after the test. Figure 3.4 shows the results. We can see that when the demographic information is asked before the exam, male performance improves, whereas female performance decreases. Asking demographic information after the test shrank sex differences in performance by 33%. If you compare this to the difference on school-based standardized tests (Figure 2.3), in which demographic information is collected completely separately from the test, you can see that the difference is much greater when collecting it before the test. While the differences appear large in this graph, the statistical significance of them is the subject of debate. There are different statistical criteria one can use to define statistical significance, and the statistical significance of this result is in between. The original paper concluded that there was no statistical significance to the differences in performance (Stricker and Ward, 2004). But given how important standardized test scores are on forming students' choices about what careers they are capable of, the lower bar for significance is merited (Danaher and Crandall, 2008).

Is there a way to neutralize stereotype threat? A real-world trial found one in the last course of an advanced university calculus sequence containing problems similar to the GRE (Good, Aronson, and Harder, 2008). All students were told the test was "aimed at measuring your mathematical abilities" (to induce stereotype threat), but half of the students were also told that "this mathematics test has not shown any gender differences in performance or mathematics ability" (to neutralize the stereotype threat). They looked at two results: the test results and course grades. For females, performance was worse than males when the stereotype threat was not neutralized, but it was the same when it was. Final grades showed no difference between males and females. In addition, females in the group in which the stereotype threat was neutralized performed better on this test than their overall course grade would predict (and females in the other group performed worse than their overall course grade would predict). This experiment has a few possible conclusions. First, stereotype threat occurs even when students are not reminded of their gender, just that the test material is testing something in which their group does stereotypically worse. Second, the threat can be neutralized. Third, females in the normal classrooms, with no intervention, suffer some degree of stereotype threat.

The effect of stereotype threat is magnified the more a group becomes a minority. When females took a math test with only other females, they performed much better than when females took the test with one male, which was better than when they took the test with two males (Inzlicht and Ben-Zeev, 2000). When females took a verbal test, the results were not affected by the gender of the other students taking the test. In addition, males' results on the math test were not affected by the gender of the other students taking the test.

Even if females made up half of the class, they would experience stereotype threat because of the stereotype that females are not as good at math. But the fact that females are such a small minority means that the effect is magnified, leading to even worse performance. There has been some notion that females tend to do better where there is *critical mass*, perhaps for social reasons. Research on stereotype threat would suggest that it is not just social—reaching critical mass allows females to perform more closely to their potential on tests.

To make matters worse, the gender of the test proctor may influence stereotype threat. Marx and Goff (2005) showed that in exams taken by African Americans and White students, African Americans performed better when the test administrator was African American, whereas there was no difference in performance of White students.

In summary, there are a number of factors that cause and magnify stereotype threat for females in computer science:

- There is a stereotype that males are better at computer science than females

- Females are a small minority in computer science classes

- Instructors and teaching assistants, and therefore the test administrators, are often male

There are several relevant effects of stereotype threat. We have only discussed test performance above, but others are just as important for advanced/honors opportunities.

- Test performance

- Blaming oneself for failure (leading to transferring out of the major)

- Self-handicapping, or not trying one's best so that there is an excuse if one does not do well.

- Changing career paths and less enthusiasm about the career.

As long as these stereotypes exist, there will be stereotype threat, leading to artificially low performance of certain groups of people. While the best solution would be to remove these stereotypes that do such harm, we focus in this book on what one individual can do. You can start by making sure you do not believe any of the stereotypes. The students you see in your course belong there because of their achievement, regardless of whatever stereotype there may be about their group. In addition, there are several small changes that professors can make to reduce the impact of stereotype threat. These are presented in the following chapters.

Further reading:

Reducing Stereotype Resources—excellent on-line resource about stereotype threat, including many more examples, explanations, and summaries of numerous experiments published in peer-reviewed publications. It provides thorough analysis of the current research, separating information into what it is, consequences, who is vulnerable, what situations lead to it, the mechanisms that cause it (which is still very much an area of research), how to reduce it, as well as criticisms about the research and unresolved issues.

`www.reducingstereotypethreat.org`

The Effort Effect

`http://alumni.stanford.edu/get/page/magazine/article/?article_id=32124`

Joshua Aronson, "Rising to the Challenge of Stereotype Threat"

`http://www.youtube.com/watch?v=ahN-dSh_1Tc`

3.2.4   DISCRIMINATION

MIT performed an equity analysis to see if female senior faculty members were being treated differently from their male counterparts. What they found was staggering. They were systematically discriminated against in small but increasing ways throughout their careers, with senior females experiencing more inequity than junior females.

How did this happen? The report states two causes. First, "Women faculty who lived the experience came to see the pattern of difference in how their male and female colleagues were treated and gradually they realized that this was discrimination. But when they spoke up, no one heard them, believing that each problem could be explained alternatively by its "special circumstances"...They found that discrimination consists of a pattern of powerful but unrecognized assumptions and attitudes that work systematically against women faculty even in the light of obvious good will" (Hopkins, 2000).

Data gathering found inequities in space, amount of nine-month salary paid from individual research grants, teaching assignments, awards and distinctions, inclusion on important committees and assignments within the department. A salary equity review in 1995 and 2000 found gender-based inequity, and salaries were adjusted as a result.

CHAPTER 4

In The Classroom Tips

The causes are social ills, and only changing society will solve them. This will require changes in the attitudes of parents, teachers, the clergy, politicians, scientists, etc. If we believe that this is the most we can do, we can become paralyzed with the enormity of the problem.

Computer science professors, while they have little power to change the first 18 years of females' lives, have many more opportunities than they might realize. We can at least retain the females that enter as freshmen. Several schools (UCSD, Harvey-Mudd, CMU, Indiana University) have found that change at the university level can work. Those changes involve every level. Here, we present the changes a single professor can make in the classroom.

For many females, the classroom is the place where professors have the most control over the females' computer science experience. While one-on-one interactions are important, large research universities do not have the luxury of developing students individually through professor-student interactions. There are, however, several ways that professors can help at the classroom level. These include what happens during lecture, how assignments are structured, and training TA's for discussion/laboratory management.

4.1 LECTURES

During lecture, professors are able to set the stage for the course. The beliefs the professor chooses to emphasize can increase or neutralize the factors contributing to stereotype threat (Section 3.2.3), instill or reduce confidence in oneself, and reinforce or reduce the challenges resulting from differences in communication patterns. Below, we have suggestions for how the professor can use these to help equalize the experience of females and males. There are four major aspects of lecture time—when the professor is talking, asking questions of students, taking questions from the students, and cooperative work.

Choose gender-neutral examples such as music, history, science, medicine, pop culture, travel, etc. rather than those that feed into existing stereotypes such as sports, cooking, sewing, video games, or science fiction. Stereotype threat is induced by reminders that one is female and is not expected to do well, which is heightened by reminders that one does not belong socially. Using examples to which females do not identify reinforce (in the females' minds) that they do not really belong.

Emphasize the class identity as college students rather than males and females. Every person belongs to multiple groups. While females are not stereotypically computer science students, females are college students, earning more than 50% of bachelor's degrees in the country. College students are known to do well in computer science. So by emphasizing everyone's equal status as a college student, this helps neutralize the stereotype threat of being female.

Emphasize the importance of practice rather than ability. Students who believe in the malleability of intelligence, the thought that intelligence is a result of practice as much as ability, perform better and are less susceptible to stereotype threat Aronson, Fried, and Good (2002).

Ask questions of females as often as males, and make sure to keep the level of difficulty the same for both. In elementary school, teachers ask boys more questions (French and French, 1984), and the types of questions are more open-ended and challenging than those asked of girls (Swann and Graddol, 1998). This leads to less confidence in answering those types of questions (Section 3.1.2, Section 3.1.1). Therefore, it is important to give as many students as possible practice in answering challenging questions during class rather than allowing a few students to control the pace of the class and receive feedback on their knowledge.

When *students answer questions*, do so in a way that instills confidence in the questioner (Section 3.1.2). For example, if the answer you hear is not what you are looking for, try to think of two things. First, in what way could my question have been interpreted to make that a correct answer? If so, acknowledge that correctness and restate the question. Second, think of some aspect of the answer that is on the right track, even if it is not actually correct.

4.2   ASSIGNMENTS

Pair programming is a very specific way of working in pairs that has been shown to increase retention (of female students especially Linda Werner et al. (2005)) as well as programming competency.

In pair programming, students do not divide the work and then work individually. Instead, two students work at the same computer. At any given time, one student is the "driver"—using the computer directly, and the other student is the "navigator"—an assistant of sorts who is supposed to be kept in the loop by the driver so that he/she can be a sounding board for ideas, ask questions when he/she is not sure, and point out potential errors. The partners switch roles often so that each can gain confidence in his/her abilities (the driver role).

When assigning pairs, it is important that students do not devolve into one student doing most of the work (or the intellectually challenging work). Unfortunately, because of gender communication patterns and social moors, when females and males are paired together, unbalanced roles are more likely to occur. Therefore, it is important to not pair females and males. In addition, when students with vastly different experience or prior achievement get paired, the stronger student is more likely to do more of the work, depriving the weaker student of the opportunity to gain experience. Therefore, it is also important to pair people who are the same or only a little different in experience.

Further Resources:

NCWIT: Pair Programming in a Box—including research behind it, full resources, and tips.

```
http://www.ncwit.org/resources/pair-programming-box-
power-collaborative-learning
```

Cluster females in group projects. Larger (3+ people) group projects are becoming increasingly popular as industry and academia requires more teamwork to create innovative solutions. Unfortunately, when females are a minority in a group (especially < 15%), it has been observed that their mistakes are amplified, they are socially isolated, and they are given roles that undermine their status (Section 3.2.1, Mcdonald, Toussaint, and Schweiger (2004)). Unless you are going to explicitly teach students about differences in communication patterns and how to detect/handle them, it is unfair to spread the females out. Instead, you should cluster females a little bit so that they are at least half of the students in a group.

If you do not assign groups, you can instead give very brief instruction on recognizing your communication style and how to work with others with different communication styles. Emphasize the importance of not interrupting, giving all students fairly equal chances to participate, brainstorming without criticism, etc. This needs to occur somewhere in the curriculum so that workplaces gradually improve, as well.

Make expectations explicit and judge by them when assigning open-ended projects. The appeal of open-ended projects is that you are providing students the opportunity for creativity and to impress you. Often, this is intended to identify students with very high creativity and knowledge. In the current world of limited time, this ends up rewarding those who are willing to sacrifice their other classes, relationships, families, etc., for this one project. Females are much less likely to be willing to do this because they are more likely to have strong family obligations, are less willing to fail in other classes, and choose safer projects (Section 3.1.1). This can be adjusted by the way in which you present and judge open-ended projects. First, give example projects that show what you consider an excellent project (that is achievable in the time provided). Make sure that you give enough time for students to achieve that level. Second, judge all projects at that level or higher as being excellent. Risk-taking behavior can be taught on an individual basis in undergraduate research or graduate school by treating failures as necessary stepping stones to success.

4.3   EXAMS

Examinations have several factors that induce stereotype threat (Section 3.2.3). First, they are diagnostic tests in a subject that females do not do well in stereotypically. Second, females are a small minority of test takers. During exams, it is important to attempt to neutralize these threats.

Remind students of gender-equal outcomes, specifically that this test has never shown any gender differences in performance or computer science ability.

Add female test administrators in introductory courses as long as this does not in any way hurt female graduate students (by assigning them to introductory courses).

4.4   LABORATORY / DISCUSSION

If you have labs manned by teaching assistants, you have an unique opportunity to monitor student-student interactions. There is what we call the *vocal minority* of experienced, macho students who brag about how little time an assignment took, ask advanced questions to show off their knowledge, and complain about how easy the class is. This is very intimidating to inexperienced students, especially those in underrepresented groups.

Train your graduate students to watch out for these comments and take the student aside. Harvey Mudd has had success with having the TA acknowledge that the student is experienced (making them feel good) but to please save those questions/refrain from those comments around others because we want to maintain a supportive atmosphere.

4.5 OFFICE HOURS

During office hours, professors have the opportunity to interact with students one on one, reacting to their individual situation. This is where communication patterns are the most damaging to females because it often involves a female student talking to a male professor (Section 3.1.3). Because of the difference in how females present themselves, it is easy for a professor to believe the student is not as capable as they are, and perhaps they should pursue other avenues.

The most important thing a professor can do is distinguish between what a student *thinks* about herself and how the student is *actually doing*. This way, it is less likely that male students will succeed in presenting themselves as more capable than they are, and less likely that female students will succeed in presenting themselves as less capable than they are.

Ask questions to gauge current knowledge before answering a question of a student. This allows the student the opportunity to guess and/or fill in the level of knowledge they have. Sometimes, the student was only needing confirmation and actually knew how to do it. Other times, the way the question is phrased leads one to believe the student is much more confused than he/she is.

Some students come to office hours with concerns about their progress, whether they are right for the major, etc. In this case, it is important to *emphasize the role of practice and effort over ability*. For any career, ability without practice or practice without ability will lead to failure. Both are necessary, yet our society has shifted to spending its time telling students they are all amazing in ability (Bronson and Merryman, 2011). The effect on this generation of the excessive praising of ability is that students are afraid to tackle difficult tasks because it might make them lose their identity as being smart (Dweck, 2007, Bronson and Merryman, 2011). In addition, once they do poorly, they are convinced they are incapable of success. Praising effort or study skills rather than intelligence or ability leads to perseverance.

Further Reading:

NCWIT: Top 10 Ways You Can Retain Students in Computing

```
https://www.ncwit.org/resources/top-10-ways-retain-
students-computing/top-10-ways-retain-students-computing
```

CHAPTER 5

Advising Tips

This chapter is for advising either graduate or undergraduate students in independent research projects or within research groups. The challenge for females is that these projects are inherently open-ended and require many more leaps from what they have seen. This is where the lack of confidence can dramatically affect the advisor's view of the student. In fact, this is what first piqued my interest in how being female affected one's career (beyond blatant discrimination).

A few years after I first started teaching, a colleague in my department approached me with the following quandary: Why did his female student have difficulty working independently on her senior project, despite her demonstrated ability in his class?

When we delved further, we discovered it was merely fear of failure and the need for reassurance, not a lack of ability, that caused her to give this impression. This led me to two questions: Had she not been in his class, how would he have recognized her potential? How can he develop in her the confidence and independence necessary to succeed at competitive levels in academia?

5.1 ONE-ON-ONE MENTORING

Initiate mentoring for female graduate students especially because of the social isolation they encounter and the new environment that is much different from their prior 16 years of schooling. For some students who were wildly successful as undergraduates, it is not until graduate school that these issues become problematic. Providing that support before it is necessary better prepares the student.

Introduce and encourage students to read about Impostor Syndrome, experienced by many students who reach levels of achievement that others do not. They feel that they do not belong there because others appear so confident. Once the students realize that a) other famously successful people had feelings of inadequacy despite their outward appearance of confidence and b) their own feelings of inadequacy are normal and not related to their future achievement, they can better suppress their doubts and concentrate on succeeding.

As a further step, *share stories of your own struggles* so that students realize that struggling is a normal part of the process, even for people who ended up being very successful. Countless times, after I've shared these stories, the students have thanked me for sharing. Many students think that they are the only ones feeling this way (especially since with so few females, they do not have the opportunity to talk to each other).

Focus on observed facts when judging students. In other words, do not believe seemingly self-aware expressions of self-doubt. Because of differences in confidence and communication patterns, females are more likely to have self-doubt, express their self-doubt, and minimize their accomplishments when speaking to others. A professor who is not used to students expressing these thoughts might erroneously think the student is amazingly self-aware and find examples that agree. This can lead the professor to pass over the student for awards, leadership positions within the research group, and lackluster letters of recommendation. We have seen this affect a graduate student's ability to get an interview at a top-tier research university.

Further Reading:

NCWIT: *How Can Encouragement Increase Persistence in Computing? Encouragement Works in Academic Settings (Case Study 1)*

```
https://www.ncwit.org/resources/how-can-encouragement-
increase-persistence-computing-encouragement-works-academic-
settings
```

5.2 ADVISING INDEPENDENT PROJECTS

Give explicit permission to fail so that students are more likely to take risks. If failure is part of the normal process, then students will fear failure less. Because students lacking in confidence are more fearful of failure, it is important to remove this barrier to tackling challenging problems.

Assign increasingly risky projects so that the student can build confidence in his/her abilities gradually.

Further Reading:

NCWIT: *How Can REUs Help Retain Female Undergraduates? Faculty Perspectives (Case Study 1)*

```
https://www.ncwit.org/resources/how-can-reus-help-retain-
female-undergraduates-faculty-perspectives-case-study-1
```

NCWIT: *REU-In-A-Box: Expanding the Pool of Computing Researchers*

```
https://www.ncwit.org/resources/reu-box-expanding-pool-
computing-researchers
```

5.3   RUNNING RESEARCH GROUPS

Provide communication training that identifies students' own communication pattern and teaches them how to relate well to students with different communication patterns.

Assign leadership rather than throwing a bunch of students together and seeing who emerges as a leader. The one who emerges as a leader is often the one who controls the conversations and pushes his/her ideas on the group. Because of the differences in confidence and communication patterns, this is much less likely to be a female than a male. Females are more likely to allow being interrupted and are less likely to interrupt males (Tannen, 1994). If you want to try different people out as leader, make this explicit, and explicitly tell the students how to act when not as leader. Training aggressive students to work well when not the leader is just as important as training timid students to be leaders.

Fund your female students to attend leadership workshops so that they can learn from other successful females and find mentors/role models.

Further Reading:

One Minute Manager by Blanchard and Johnson.

CHAPTER 6

Faculty Recruiting / Retention Tips

Female faculty members face unique challenges at all stages in their academic careers: graduate school through the hiring process, as nontenured faculty members, and advancing to full professor. The root cause is the full transition from objective exams in which everyone answers the same questions (and to which there are correct answers) to evaluating creative, unstructured research endeavors that are inherently hard to compare.

Theoretically, the candidate should be judged only on the quantity and quality of publications, along with the person's contribution to those papers (i.e., first author vs later author), as well as their vision expressed in application documents. If hired, all faculty should be given the same opportunities to succeed.

The reality appears to be far from this ideal. There is continuing sexism in the peer-review process when judging based on these documents, as well as differences in the ways women are viewed by those who work closely with them (resulting in weaker letters of recommendation). In addition, females' socialization against negotiation places females at a disadvantage for their entire careers. Finally, there remains a great imbalance in both the roles females and males take in producing and rearing children and the way others view them for becoming parents.

6.1 FACULTY RECRUITING COMMITTEES

These problems first appear in and to the faculty recruiting committees, but they continue through promotion cases (especially letter cases). This is a case where changing one person's actions is not enough—the entire committee or department needs to be made aware of these issues.

Inform faculty evaluating current and potential faculty as to these effects. Certainly, case leads and the committee chairs need to have read about biases and how to lead committees to reduce their effects. In a committee, the chair could then present the findings and tips to the rest. At the beginning of the faculty evaluation "season," the chair could similarly present this information to the faculty. Ground rules about discussions and evaluations could be established at this time.

Further Reading:

Unconscious Schemas: Bias and Assumptions—University of Missouri Equity Office

http://equity.missouri.edu/recruitment-hiring/bias.php

WISELI, the Women in Science and Engineering Leadership Institute at the University of Wisconsin-Madison, including two relevant publications:

Searching for Excellence and Diversity: A Guide for Search Committee Chairs, Eve Fine and Jo Handelsman (University of Wisconsin-Madison, 2005)

Reviewing Applicants: Research on Bias and Assumptions by Eve Fine and Jo Handelsman (University of Wisconsin-Madison, 2006)

6.2   SEXISM IN PEER-REVIEW

People have a relatively strong belief that their actions, including evaluations of a set of applicants based on their academic record, are based on conscious thought. Most faculty, if asked, would probably say that they judge people fairly, regardless of their gender, ethnicity, and connections to that person. Unfortunately, research shows that implicit biases affect us much more than we would like.

Wenneras and Wold (1997) performed a through analysis of the peer-review system of the Swedish Medical Research Council (MRC), one of the main funding agencies for bio-medical research in Sweden. Although there is gender balance in the number of applicants each year, many more grants are awarded to males. First, the authors formulated an equation for scholarly impact based on the quantity and quality of publications, first-author status, and number of citations. They found that, within a gender, the scholarly competence score of the reviewers roughly followed their scholarly impact score. They found, though, that only the women with the highest total impact were rated as high as any males—and only as high as the males with the lowest research productivity.

To delve further, they performed regression analysis to determine which factors (scholarly impact, gender, affiliation with a committee member, nationality, type of Ph.D., university affiliation, evaluation committee, postdoc experience, and a letter of recommendation) correlated with the scientific competence score given by the reviewers. They found that affiliation with a committee member *or* being male provided a boost equivalent to, in their calculations, three extra papers in *Na-*

ture or *Science*, or 20 extra papers in a journal with fairly high impact. A female with no affiliation with a committee member had virtually no chance to obtain funding, since they would require twice this *beyond their original productivity* to be scored equally with a male affiliated with a committee member.

There have been many studies showing that, without intervention, people unconsciously give scores that reflect who the person is rather than what the person did (regardless of the gender of the evaluator). Because peer-review evaluations are the main method of evaluating academics, females face these disadvantages at all stages of their career. Faculty recruiting committees have the same information and produce similar subjective measures, NSF review panels have an explicit rating for the researcher themselves, and all merits and reviews are performed in a similar manner. This affects females' ability to get interviews, job offers, funding, tenure, and full professor.

Inform peer-review committees of such biases and instruct them to make sure their scores are based on *achievement* rather than *perception*.

Create rough equations to calculate the scholarly impact that take into account quantity, quality, author position, and citations of scholarly works. Provide these with each record. While not binding, these would likely cause committee members to take a second look when their scores do not match the scholarly impact score.

Double-check scores, looking at the reasons the scholarly impact scores did not match the committee members' rankings to make sure a systemic bias is not being introduced (i.e., gender, connections, or ethnicity).

Make double-blind reviews whenever possible. For example, for NSF reviews, the CV could be scored separately from the meat of the research proposal. All conferences should employ double-blind review. For department recruiting committees, and promotion committees, this may be impossible since the record, not a research idea, is what is being evaluated.

6.3   LETTERS OF RECOMMENDATION

There are two challenges females face in letters of recommendation. The first is that the adjectives used in letters for females (communal adjectives) are very different from the adjectives used in letters for males (agentic adjectives) (Madera, Hebl, and Martin, 2009). The second, and very related, problem is that the communal adjectives (used more often for females), even when separated from gender, were negatively related to hireability ratings. Table 6.1 gives examples of

Table 6.1: Examples of communal and agentic adjectives, followed by proposed adjectives and positive skills. Source: Madera, Hebl, and Martin (2009)

Communal Adjectives	affectionate, helpful, kind, sensitive, agreeable, tactful, sensitive, nurturing, warm, caring
Agentic Adjectives	assertive, confident, aggressive, ambitious, dominant, independent, daring, outspoken, intellectual
Positive Adjectives	creative, intelligent, talented, brilliant, analytical, rigorous, broad, deep
Positive Skills	excellent speaker, leader, writer, perspective, ability to choose timely problems

communal and agentic adjectives. This is related to the Gender Roles Obstacle (Section 3.2.2).

There are two separate effects here. The first is possible sexism—that women who achieve at the same level of men are described differently in letters of recommendation, making them less likely to get hired or obtain funding. The second is a possible difference in the value the letter writers and/or committee placed on certain attributes.

The Madera, Hebl, and Martin (2009) study supports both possibilities. There does appear to be sexism, because gender had a much stronger correlation to the presence of communal and agentic adjectives than any performance measures such as years in graduate school, number of publications, number of first-author publications, honors, years taught, and position applied for. Therefore, it appears that regardless of one's *achievements*, the letters of recommendation are colored by one's gender.

On the surface, though, a more surprising result involved the gender of the letter writers. Female letter-writers were more likely than males to describe females in communal terms and males in agentic terms. If sexism is the only issue, then we would expect male letter-writers to use the apparently less valuable attributes.

Perhaps because more than sexism is involved. Females are socialized from an early age to *be* more communal, starting with preschool and segregated toy sections in toy stores (see Section 2.3 for more details). If this is true, then it is in the *interpretation* of these adjectives that the trouble might lie.

For example, one could argue that being aggressive, dominant, and outspoken might not be better traits for successful research than being agreeable, tactful, and helpful. The former are more likely to burn bridges, hurting their chances of collaborative research. One could easily argue that the future of research is in interdisciplinary research, and *nice* people who are also intelligent and creative are more likely to lead groups that make innovations there. In fact, the results in the collective intelligence study by Wolley et al. (2010) showed that *being female* and *social intelligence* correlated with group success, but individual intelligence did not.

The key is that when you are using communal adjectives, because they have a negative connotation (sometimes meaning the lack of other positive attributes), make sure that you convey to the letter-writer that you are meaning them as a positive. This can be done by first emphasizing the presence of other positive attributes (i.e., creativity, leadership, intelligence) with the enhancement of more (collaborative, rigorous, great communication skills). More specific recommendations are below.

6.3.1 WRITING LETTERS

Letter-writers need to be careful not to use communal adjectives blithely. They need to be meaningful, coupled with concrete positive results, and preceded by more strongly positive attributes (i.e., creativity).

Choose gender-neutral, positive adjectives and fit them with every candidate, regardless of gender. Especially if you are female, realize that the adjectives you use, and perhaps even value, are not necessarily read in the light that you wrote them.

Go beyond adjectives, illustrating how those attributes have or you believe will have a direct, positive effect on that person's career.

Order matters—if a candidate is collaborative, that is okay to say, but be aware that this is a word that can fall in the communal category. If someone is only collaborative, then it may imply that someone has a weakness in creativity or leadership. But if you have already made the case for creativity and leadership, collaborative is a positive.

Read letters of recommendation of faculty candidates so that you are learn which adjectives the strong candidates' letter-writers use.

Provide explicit opportunities for students to display the attributes that result in strong letters of recommendation.

6.3.2 READING LETTERS

More than anything, letter readers need knowledge and perspective.

Always be aware of gendered adjectives for female candidates so that you don't allow the letter-writer's gender stereotypes to negatively affect your view of the candidate.

Focus on quantitative measures to judge the promise of the candidate.

Expand your view of a successful candidate to include those who have communal attributes. Someone who is successful and *nice* might be a better candidate than one who is equally successful but burns bridges.

6.4   CULTURE OF NEGOTIATION

As with many jobs, salary is negotiated for faculty positions, and differences in willingness and/or success in negotiating may lead to a gender pay gap. The large inequity is that the negotiations also include items crucial to professional success such as start-up money to pay graduate students prior to obtaining grants and how much lab space you have for students and equipment. Recent research has shown that females are denied their requests more often than men (Section 3.2.2, Catalyst (2011)).

In order to counteract the unfairness present in the negotiation process, it is important to conduct *regular parity reviews* of critical resources to make sure they are being distributed fairly.

Conduct regular parity reviews on areas such as space and salary, making sure those are commensurate with *performance* rather than negotiation.

Apply competing offers to all rather than to only the person who received it to avoid a chasm between people willing to get competing offers and those not willing to.

6.5   N-BODY PROBLEMS

Females often face greater challenges than males at two critical junctures—finding a position while married and raising children. A comprehensive study was performed by the Stanford Clayman Institute for Gender Research (Schiebinger, Henderson, and Gilmartin, 2008). The results show the magnitude of this problem for females.

The first problem, typically called the two-body problem, occurs when a female looking for a position takes into account what job offer her partner receives when deciding the position. This is a larger problem for females in male-dominated fields because females are much more likely to be coupled with a male in the same

Table 6.2: Percentage of faculty with academic spouses in the same department, broken down by gender and field, ordered by the ratio of female percentage over male percentage. Source: Schiebinger, Henderson, and Gilmartin (2008)

Field	Male	Female	Ratio
Computer Science	11.1%	44.4%	4
Engineering	24.7%	64.3%	2.6
Law	38.4%	78.9%	2.1
Business	27.8%	48.4%	1.7
Science	54.2%	82.7%	1.5
Medical	70.3%	68.3%	0.97
Humanities	78.0%	69.6%	0.89
Education	54.5%	35.9%	0.66

field. While 40% of female faculty and 34% of male faculty have an academic partner, the rates are much higher in STEM. Table 6.2 shows the percentage of females with partners in the same department versus males, for each field. We can see that in computer science, females are four times as likely to have a spouse in the same department as males. In female-dominated departments such as education, the gap is reversed. In addition, within each rank, females were approximately half as likely as men to report that their career was a priority over their spouse's career.

When searching for a position, the need to find a job for a spouse further erodes the negotiating position of the female. In addition, the number one reason women refused an outside offer was because their academic partners were not offered appropriate employment at the new location.

When couples have children, despite advances in the past century in the role of women, the brunt of the child-rearing still falls on the females. Current studies suggest breast-feeding for at least a year, something that males are clearly unable to do for their wives.

For the time period 2005-2009, only half of full-time employed married men took care of their kids, and those who did spent only around 52 minutes per day doing so. In contrast, 72% of full-time employed married women, took care of their children, spending an average of 73 minutes per day taking care of them. This is not offset by household tasks by men—men reported having at least a half hour more time than women to spend on "relaxing and leisure" (Labor Department, 2011).

This inequity hurts females' research and networking output, limiting their time with graduate students and travel to conferences and meetings.

The university can help females during both of these times.

Proactively help find spouses jobs so that females do not have to turn down positions due to lack of appropriate employment for his/her spouse.

Do not punish candidate for spouse job-hunting—perform normal negotiations separate from the spouse's job.

Stop the tenure clock as a formal policy so that females have time to care for newborns. Remember that when a clock is stopped, when the person goes up, the record is compared against someone who had fewer years to build their record. Females should feel that this is the normal route, not an exception that will hurt them later.

Provide high-quality, low-cost infant/child daycare so that female faculty members can focus on their work after maternity leave.

Provide females teaching release beyond maternity leave so that they can put their little free time toward research rather than teaching. Research programs cannot go on hiatus, whereas a gap in teaching will not impact a long-term career.

Further Reading:

Does 'Leaning In' Actually Work for Women at the Starting Line—Three women, including the CEO of Piazza, weigh in on Sheryl Sandberg's new book.

http://www.npr.org/2013/04/03/176130221/does-leaning-in-actually-work-for-women-at-the-starting-line

What men can learn from Sheryl Sandberg's 'Lean In'—Sheryl Sandberg and AmEx CEO Ken Chenault

http://www.youtube.com/watch?v=9TjRVDUERY4

CHAPTER 7

Institutional Change Tips

While individuals can make large impacts on small numbers of students, the changes that make the most impact require change at the department or university level. In fact, *historically, the successful programs have required changes at all levels, not just faculty*. We do not want to dissuade individual faculty members from doing their part—the impact that individual professors can make on individual students cannot be emphasized enough. However, *to make large-scale changes, institutional changes are necessary*. We encourage readers to learn about success stories at Carnegie Melon University, Harvey Mudd, and others. Revolutionary change will only occur through cultural instutional changes.

7.1 DEPARTMENT

The department has the largest control over the experience for undergraduate students, since they do not interact as closely with faculty as graduate students. Prior success stories at Carnegie Melon University and Harvey Mudd University have shown what a department-level commitment can make.

These suggestions can be taken piecemeal or as part of a large goal: *Make interdisciplinary work a part of the culture and deemphasize "hacking for the sake of hacking."*

This is related to a previous tip about open-ended problems (4.2) that reward students willing to prioritize hacking over other classrooms, family life, and all other interests. As long as the culture (and reward system) exists that only those who absolutely love and will dedicate themselves to hacking will excel, we will continue to attract the narrow sliver of the population who believes this. This is perhaps the most challenging problem because many of the faculty succeeded because of their belief or acclimation to this culture.

Partition by experience in the first class so that inexperienced students are not intimidated by the experienced students. Females are overly represented in the inexperienced students, so it is important that those students have a positive environment to build their skills and confidence.

Train faculty and teaching assistants to diffuse macho culture. There exists some "vocal minority" machismo in which students ask questions to show their knowledge, brag about how short a time an assignment took, and how little time they studied. These tendencies have no positive outcome because they perpetrate two incorrect views that hurt females: knowledge at the freshman level and the lack of studying have a positive correlation to success. Studies on praise (Dweck, 2007) and stereotype threat (Aronson, Fried, and Good, 2002) have shown that believing in effort rather than current knowledge has a positive effect on success. Therefore, both of these attitudes are highly corrosive and should be minimized.

Do not filter too early. When computer science departments must turn students away, they make standards for transferring into the major more stringent, increasing GPA requirements on a certain set of courses. It is important to have at least a year's worth of courses because if you filter too early, you are filtering for experience rather than achievement. This also filters females, since they are less likely to have experience.

Reduce stereotype threat by scheduling female teaching assistants in the introductory courses (as long as this has no negative effect on the graduate students).

Provide female role models by taking care to have computer science majors encounter at least one female faculty member early in their career, as well as female teaching assistants. Invite female speakers to give presentations to graduate students and faculty.

Ensure that females are invited for guest lectures to provide role models and examples for female and male students and faculty.

Sponsor a female-oriented club so that females can feel a sense of belonging with peers. To elevate their status, encourage them to run important opportunities like: informal advising session before signing up for classes.

Encourage faculty to read this book or provide other training to prevent flagrant violations or unintentional bias within the department. One department we know of (which will remain nameless) denied tenure to a female who stopped her clock, judging her based on the longer timeframe rather than the normal timeframe. She successfully sued the university. She won because her case was so clear cut, but others may not be so lucky. Prevention is better than correction or punishment.

7.2   UNIVERSITY

Many of the initiatives for faculty require university-level approval. University-sponsored daycare, regular equity analysis of salaries, lab space, and start-up packages, stopping the tenure clock, teaching releases, and spousal hires require approval above the department.

More than anything, the university needs to emphasize that equity is an institutional goal for which it is willing to provide resources. These resources might include direct funding for an on-campus childcare facility, "spouse slots" that do not count against a department's growth plan for five years, and funding for lecturers to cover teaching releases.

Further Reading:

NCWIT: How Does Engaging Curriculum Attract Students to Computing? Harvey Mudd College's Successful Systemic Approach (Case Study 2)

```
https://www.ncwit.org/resources/how-does-engaging-
curriculum-attract-students-computing-harvey-mudd-colleges-
successful
```

Bibliography

Carol Dweck, *Mindset: The New Psychology of Success*, Ballantine Books, 2007. 45, 60

Bronson and Merryman, *NurtureShock: New Thinking About Children*, Twelve, 2011. 45

K. Bennhold, "Toys Start the Gender Equality Rift," *The New York Times* July 31, 2012. 13

M. Tannenbaum, "The Problem When Sexism Just Sounds So Darn Friendly…," *Scientific American Blogs*, April 2, 2013. 34

Moss-Racusin, "Disruptions in Women's Self-Promotion: The Backlash Avoidance Model," *Psychology of Women Quarterly* vol. 34, no. 2, pp. 186–202, 2010. DOI: 10.1111/j.1471-6402.2010.01561.x

Williams and Dempsey, "Women's Career Advice: Self-Promote without the Backlash," *Huffington Post Blog*, October 14, 2011. 35

Steven Jay Gould, *The Mismeasure of Man*, W. W. Norton and Co., 1981, 1996. 16

Taulbee, "CRA Taulbee Survey," Computing Research Association, 2010. 32

Gerdes and Gransmark, "Strategic Behavior across Gender: A Comparison of Female and Male Expert Chess Players," *IZA Discussion Papers* 4793, Institute for the Study of Labor(IZA), 2010. DOI: 10.1016/j.labeco.2010.04.013 11

Kanter, *Men and women of the corporation.* New York: Basic Books, 1977. 33, 34

Mcdonald et al., "The influence of social status on token women leaders' expectations about leading male-dominated groups," *Sex Roles* vol. 50 pp. 401–409, 2004. DOI: 10.1023/B:SERS.0000018894.96308.52 33, 34, 43

Glick and Fiske, "The Ambivalent Sexism Inventory: Differentiating Hostile and Benevolent Sexism," *Journal of Personality and Social Psychology*, vol. 70, no. 3, pp. 491–512, 1996. DOI: 10.1037/0022-3514.70.3.491 33

Tannen, "The relativity of linguistic strategies: rethinking power and solidarity in gender and dominance," *Gender and Discourse*. Oxford University Press, Oxford and New York, pp. 19–52, 1994. 29, 32, 33, 49

S. S. Case, "Gender differences in communication and behaviour in organizations," *Women in Management: Current Research Issues*. pp. 144–67. 33

Schmid-Mast, "Gender differences and similarities in dominance hierarchies in same-sex groups based on speaking time," *Sex Roles*, vol. 44, pp. 537–556, 2001. DOI: 10.1023/A:1012239024732

Smith-Lovin, "Interruptions in Group Discussions: The Effects of Gender and Group Composition," *American Sociological Review*, vol. 54, pp. 424–435, June, 1989. DOI: 10.2307/2095614 33

Catalyst, "The Myth of the Ideal Worker: Does Doing All the Right Things Really Get Women Ahead?" *Catalyst Knowledge Center Research Report*, October 1, 2011. 34, 56

Heilamn et al., "Penalties for success: Reactions to women who succeed at male tasks," *Journal of Applied Psychology*, vol. 89, pp. 416–427, 2004. DOI: 10.1037/0021-9010.89.3.416 35

Zahn-Waxler et al., "Guilt and empathy: Sex differences and implications for the development of depression," *The development of emotion regulation and dysregulation*, pp. 243-272, Cambridge University Press, 1991. DOI: 10.1017/CBO9780511663963.012 29

Brody, "Gender, emotion, and expression," *Handbook of emotions*, New York: Guilford Press, 2000. 29

Byrns et al., "Gender differences in risk taking: A meta-analysis," *Psychological Bulletin*, vol. 125, no. 3, pp. 367–383, May 1999. DOI: 10.1037/0033-2909.125.3.367 12, 28

Jianakoplos and Bernasek, "Are Women more Risk Averse?" *Economic Inquiry*, vol. 36, iss. 4, pp. 620–630, October 1998. DOI: 10.1111/j.1465-7295.1998.tb01740.x 11

C. Emily Feistritzer, "Profile of Teachers in the U.S. 2011," *National Center for Education Information*, 2011. 6, 32

NCES, *National Center for Education Statistics*, 2000. 5, 9

Steven Chu, "The Nobel Prize in Physics 1997, Steven Chu Autobiography," `http://www.nobelprize.org/nobel_prizes/physics/laureates/1997/chu-autobio.html`. 13, 18

USDOL, *2011 American Time Use Survey*, United States Department of Labor, 2011. 57

Schiebinger et al., "Dual-Career Academic Couples: What Universities Need to Know," *Michelle Clyman Institute for Gender Research*, Stanford University, 2008. 56, 57

Nora Newcombe, "Spatial Skills and Success in STEM: Thinking about Gender Differences," *NCWIT 2012 Summit*, `http://vimeo.com/channels/372194/45873134`. 17, 18

Volvo Press Release, "YCC | Your Concept Car - by Women", 2004, `http://www.volvoclub.org.uk/press/pdf/presskits/YCCPressKit.pdf`. 19

Wenneras and Wold, "Nepotism and sexism in peer-review," *Nature*, vol. 387, pp. 341–343, 1997. DOI: 10.1038/387341a0 52

Madera et al., "Gender and Letters of Recommendation for Academia: Agentic and Communal Differences," *Journal of Applied Psychology*, vol. 94, no. 6, pp. 1591–1599, 2009. DOI: 10.1037/a0016539 53, 54

Patterson and Trasti, *Women Students in Computer Science: Student Perspectives of Faculty Bias as a Possible Influence on Student Retention* Mid-Atlantic Conference on the Scholarship of Diversity. 29

Werner et al., "Want to Increase Retention of Your Female Students?" *Computing Research News*, vol. 17, no. 2, 2005. 42

Hill et al., "Why So Few? Women in Science, Technology, Engineering, and Mathematics," *American Association of University Women*, 2010.

Steele and Aronson, "Stereotype threat and the intellectual test performance of African Americans," *Journal of Personality and Social Psychology*, vol. 69, no. 5, pp. 797–811, 1995. DOI: 10.1037/0022-3514.69.5.797 35, 36

Inzlicht and Ben-Zeev, "A threatening intellectual environment: Why females are susceptible to experiencing problem-solving deficits in the presence of males," *Psychological Science*, vol. 11, pp. 365-371, 2000. DOI: 10.1111/1467-9280.00272 36, 38

Stricker and Ward, "Stereotype Threat, Inquiring About Test Takers' Ethnicity and Gender, and Standardized Test Performance," *Journal of Applied Social Psychology*, vol. 34, no. 4, pp. 665–693, 2004. DOI: 10.1111/j.1559-1816.2004.tb02564.x 37

Danaher and Crandall, "Stereotype Threat in Applied Settings Re-Examined," *Journal of Applied Social Psychology*, vol. 38, no. 6, pp. 1639–1655, 2008. DOI: 10.1111/j.1559-1816.2008.00362.x 37, 38

Schamder and Johns, "Converging Evidence That Stereotype Threat Reduces Working Memory Capacity," *Journal of Personality and Social Psychology*, vol. 85, pp. 440–452. DOI: 10.1037/0022-3514.85.3.440 36

Good et al., " Problems in the pipeline: Stereotype threat and women's achievement in high-level math courses," *Journal of Applied Developmental Psychology*, vol. 29, no. 1, pp. 17–28, 2008. DOI: 10.1016/j.appdev.2007.10.004 36, 38

Gunderson et al., "Parent Praise to 1- to 3-Year-Olds Predicts Children's Motivational Frameworks 5 Years Later," *Child Decelopment*, online, February 11, 2013. DOI: 10.1111/cdev.12064 11, 12

Spencer et al., "Stereotype Threat and Women's Math Performance," *Journal of Experimental Social Psychology*, vol. 35, pp. 4–28, 1999. DOI: 10.1006/jesp.1998.1373 36

Aronson et al., "When white men can't do math: Necessary and sufficient factors in stereotype threat," *Journal of Experimental Social Psychology*, vol. 35, pp. 29–46, 1999. DOI: 10.1006/jesp.1998.1371 36

Aronson et al., "Reducing the Effects of Stereotype Threat on African American College Students by shaping theories of intelligence," *Journal of Experimental Social Psychology*. vol. 38, pp. 113–125, 2002. DOI: 10.1006/jesp.2001.1491 42, 60

Marx and Goff, "Clearing the air: The effect of experimenter race on target's test performance and subjective experience," *British Journal of Social Psychology*, vol. 44, no. 4, pp. 645–657, 2005. DOI: 10.1348/014466604X17948 38

French and French, "Gender imbalances in the primary classroom: an interactional account," *Educational Research*, vol. 26, no. 3, pp. 127–136, 1984. DOI: 10.1080/0013188840260209 14, 42

Swann and Graddol, "Gender inequalities in classroom talk," *English in Education*, vol. 22, pp. 48–65, 1998. 14, 42

Erhardt et al., "Board of Director Diversity and Firm Financial Performance," *Corporate Governance: An International Review*, vol. 11, iss. 2, pp. 102–111, April 2003. DOI: 10.1111/1467-8683.00011 22

Groysberg, "Chasing Stars: The Myth of Talent and the Portability of Performance," 2010. 22

van Buskurk, "How the Netflix Prize Was Won," *Wired*, September 22, 2009. 20

Wolley et al., "Evidence for a collective intelligence factor in the performance of human groups," *Science*, vol. 330, pp. 686-688 DOI: 10.1126/science.1193147 21, 55

Margolis and Fisher, *Unlocking the Clubhouse: Women in Computing*, MIT Press, 2003. 13, 19, 29

Hopkins, " An MIT Report on the Status of Women Faculty in Science Leads to New Initiatives to Increase Faculty Diversity," *MIT Faculty Newsletter*, October 2000. 40

Block and Robins, "A Longitudinal Study of Consistency and Change in Self-Esteem from Early Adolescence to Early Adulthood," *Child Development*, vol. 64, pp. 909–923, 1993. DOI: 10.2307/1131226 14, 15

Stancey and Turner, "Close women, distant men: Line bisection reveals sex-dimorphic patterns of visuomotor performance in near and far space," *British Journal of Psychology*, vol. 101, no. 2, pp. 293–309, 2010. Article first published online : 24 DEC 2010 DOI: 10.1348/000712609X463679 16

Confer et al., "Evolutional Psychology: Controversies, Questions, Prospects, and Limitations", *American Psychologist*, vol. 64, no. 2, pp. 110–126, 2010. DOI: 10.1037/a0018413

Silverman et al., "Testosterone levels and spatial ability in men," *Psychoneuroen-docrinology*, vol. 24, no. 8, pp. 813–822, 1999.
DOI: 10.1016/S0306-4530(99)00031-1 17

Moffat and Hampson, "A curvilinear relationship between testosterone and spatial cognition in humans: Possible influence of hand preference," *Psychoneuroen-docrinology*, vol. 21, iss. 3, pp. 323–337, April 1996.
DOI: 10.1016/0306-4530(95)00051-8 17

Elizabeth Hampson, "Estrogen-related variations in human spatial and articulatory-motor skills," *Psychoneuroendocrinology*, vol. 15, iss. 2, pp. 97–111, 1990. DOI: 10.1016/0306-4530(90)90018-5 17

Author's Biography

DIANA FRANKLIN

Diana Franklin is tenured teaching faculty and Director of the Center for Computing Education and Diversity at UC Santa Barbara. Franklin received her Ph.D. from UC Davis in 2002. She is a recipient of the NSF CAREER award and an inaugural recipient of the NCWIT faculty mentoring award. She was an assistant professor (2002-2007) and associate professor (2007) of Computer Science at the California Polytechnic State University, San Luis Obispo, during which she held the Forbes Chair (2002-2007). Her research interests include parallel programming and architecture, computing education, and ethnic and gender diversity in computing.

Printed in the United States
by Baker & Taylor Publisher Services